CW00411273

Michael Newgass

# Human Chronology

an Historical Framework
from an English Perspective

Human Chronology V3.0

First published in 2015 by
Arnison Newgass
Gambledown Farm
Sherfield English
Romsey SO51 6JU
United Kingdom

© Michael Newgass 2015

ISBN 978-0-9541005-2-0

Cover photograph:
Silbury Hill, Avebury, Wiltshire, 2400 BCE

# INTRODUCTION

Time is linear. To make sense of the past, we need to know the order of events – what happened first and what next. The time lapse between them is important too. Simple dates solve both those problems. The purpose of this small book is to provide a framework of such dates. It cannot begin to be a comprehensive schedule and it is not a history lesson. Rather it is a partial view, seen from the vantage point of England in the year 2014. We are all the creatures of our history and geography: So, to some degree this is a personal perspective, detailing those events that shape my own world view and sense of identity.

Living just in this present moment may be an ideal, but it is an unattainable aim. Only those who have completely lost their memory can truly live in the moment. For the rest of us, we do live in the present, but our context is all that has gone before – everything we know. What follows therefore is a chronology of events that helps us to understand how we have arrived in the here and now.

The chronology is divided into arbitrary eras for ease of reference. In reality the different periods of history overlap and merge into each other. Time accelerates, or at any rate that is our experience of it. Also more recent events generally have more importance to the present. So, for example, the list of Twentieth Century events is much denser than the Eighteenth. The further back we look, the sparser are the events we can discern, and beyond a certain point their dates become uncertain.

Different threads run through this list: Natural, environmental, agricultural, artistic, religious, scientific and technological. There are commercial and military threads and inevitably some are political. The threads are such that many events attach to more than one of them, but through it all they bind together into a ribbon of human consciousness. Everything is connected.

So please use this framework to weave in those moments of our history that you come across, those that seem significant to you, so as to put them into context.

# Human Chronology
## an Historical Framework from an English Perspective

| Ga | Planet Earth *(Ga = Billions of years ago)* |
|---|---|

| | |
|---|---|
| 13.80 | Age of the Universe ('Big Bang') |
| 13.20 | Age of our Galaxy (*i.e.* of oldest known object in the Milky Way) |
| 4.50 | Age of Earth (today probably half way through its life) |
| 3.40 | Bacteria appeared (fossilised Cyanobacteria, Western Australia) |
| 2.40 | 'Great Oxygenation Event' (from Cyanobactera photosynthesis) |
| 2.00 | Eukaryotic cells appeared (nucleus and membrane, building block of all complex life) |
| 1.70 | Mitochondria arose (probably from bacteria by endosymbiosis) uncertain date |
| 1.00 | Chloropolasts arose (from an endosymbiotic event) very uncertain date |

| Ma | Evolution of Life on Earth *(Ma = Millions of years ago)* |
|---|---|

| | |
|---|---|
| 760.00 | Fungal organisms appeared (probably colonized the land by 500 Ma) |
| 650.00 | 'Snowball Earth' Event, ended by global volcanic activity (late Pre-Cambrian) |
| 570.00 | *Charnia* fractal fossils (early multicellular organism, Charnwood Forest, 1957 CE) |
| 542.00 | 'Cambrian Explosion' (Burgess Shale 505 Ma, found by Charles Walcott 1909 CE) |
| 450.00 | Ordovician-Silurian Extinction Event (global cooling & lowering sea level) |
| 395.00 | Earliest known Tetrapod (tracks found in sediment, Świętokrzyskie Mountains) |
| 360.00 | Late Devonian Extinction Event (perhaps a series over several millions of years) |
| 350.00 | Ferns became the dominant land plants (Carboniferous 359 to 299 Ma approx.) |
| 250.00 | Permian-Triassic Extinction Event (Siberian Traps, $CO_2$, acid rain, 12°C rise) |
| 230.00 | Dinosaurs appeared (later Sauropods weighed up to 80 tonnes) |
| 200.00 | Triassic-Jurassic Extinction Event (50% of species on Earth disappeared) |
| 175.00 | Pangea Supercontinent began to break up into what became modern continents |
| 150.00 | *Archaeopteryx* fossil (intermediate between dinosaurs and modern birds) |
| 100.00 | Flowering Plants appeared (seed formation and bees preserved in amber, Burma) |
| 66.00 | Cretaceous-Palaeogene Extinction Event (K-Pg alias K-T, Chicxulub Crater) |
| 60.00 | North American and Eurasian continents started to separate and drift apart |
| 47.00 | *Darwinius Masillae* fossil (Lemur like 'Ida,' Messel 1983, re-evaluated 2007) |
| 35.00 | India started to collide with Asia forming the Himalayan range |
| 5.30 | Mediterranean Sea formed following subsidence of ridge at the Straits of Gibraltar |
| 3.30 | Earliest known Stone Tools (Lake Turkana, Rift Valley, Northern Kenya) |

| Ma | |
|---|---|
| 3.20 | *Australopithecus Afarensis* ('Lucy' fossil, Ethiopia, found 1974 CE) |
| 2.60 | Oldowan Stone Tools (Gona, Afar Triangle, Ethiopia) |
| 2.60 | Pleistocene started (repeated glaciations up to the end of Younger Dryas 11,700 BP) |
| 2.50 | *Australopithecus Africanus* ('Taung Child' fossil, South Africa, found 1924 CE) |
| 1.98 | *Australopithecus Sediba* fossil (Malapa Cave, Gauteng, South Africa) |
| 1.80 | *Homo Habilis* (OH 24 'Twiggy' fossil, Olduvai Gorge, Tanzania, found 1968 CE) |
| 1.80 | Earliest humans found outwith Africa, *Homo Erectus Georgicus* (Dmanisi, Georgia) |
| 0.80 | Human footprints (*Homo Antecessor* or *Heidelbergensis*, Norfolk coast, England) |
| 0.50 | *Homo Heidelbergensis* fossils ('Heidelburg Man' Germany, 'Boxgrove Man' UK) |
| 0.40 | Use of fire at Beeches Pit, Suffolk (but fire perhaps 0.79 Ma in Israel) |
| 0.25 | Fire Hearth at Harnham, Wiltshire (charcoal and hand axes, but perhaps 0.30 Ma) |
| 0.25 | *Homo Neanderthalensis* (Pontynewydd, Denbighshire, UK) |

| BP | Human Evolution *(BP = years before present)* |
|---|---|
| 195,000 | *Homo Sapiens* ('Omo I' fossil, Kibish, Ethiopia, found 1967 CE by Richard Leakey) |
| 170,000 | *Homo Sapiens* earliest occupation of Pinnacle Point Cave, South Africa |
| 125,000 | Previous warmest Interglacial peak (Eemian, forested up to the North Cape) |
| 95,000 | *Homo Sapiens* ('Qafzeh VI' fossil, Israel) early migration died out *ca.* 80,000 BP |
| 73,000 | Lake Toba Super Eruption (est. 2,000+ km³, 4°C Earth surface temp. fall) ± 4,000 years |
| 70,000 | Engraved Ochre abstract design (Blombos Cave, South Africa) ± 5,000 years |
| 66,000 | *Homo Sapiens* earliest successful migration out of Africa (via Red Sea Route) |
| 64,000 | Stone Arrow, Dart or Spear Points (Sibudu Cave, South Africa) |
| 60,000 | *Homo Neanderthalensis* burials, incl. disabled man (Shanidar Cave, Zagros Mts) |
| 50,000 | *Homo Neanderthalensis,* including evidence of cannibalism (El Sidrón Cave, Spain) |
| 45,000 | *Homo Sapiens*, Anatomically Modern Humans reached Europe, nominal date |
| 44,000 | Hunting Artefacts, incl. poison arrows *per* San culture (Border Cave, South Africa) |
| 40,000 | *Homo Neanderthalensis* ('Neanderthal I', Feldhofer Cave, Germany, found 1856 CE) |
| 40,000 | *Homo Sapiens* ('Mungo Man' burial, Australia) ± 10,000 years |
| 40,000 | Denisova Cave Hominin (common ancestry with Neanderthals *per* mtDNA, Siberia) |
| 40,000 | European Cave Art (hand stencils and discs, Cueva de El Castillo, Spain) |
| 40,000 | South East Asia Cave Art (hand stencils, Sulawesi, Indonesia) |
| 40,000 | Zoomorphic Ivory Sculpture (Lion Headed Man, Stadel Cave, Germany) |
| 39,000 | Heinrich Event 4 (North Atlantic suddenly cooled 5-10°C, demise of Neanderthals) |
| 36,000 | European Paintings (Chauvet Cave, Ardêche, France, cave blocked 23,000 BP) |

| BP | |
|---|---|
| 35,000 | South East Asia Cave Painting (figurative painting of a Pig, Sulawesi, Indonesia) |
| 35,000 | Examples of Flutes appear (Hohle Fels cavern, Swabian Alps, Southern Germany) |
| 35,000 | Aurignacian Human Figurine ('Venus' of Hohle Fels) but perhaps 40,000 BP |
| 35,000 | Sewing Needles appear in Upper Paleolithic sites in Europe |
| 33,000 | European Ritual Burial ('Red Lady of Paviland', Gower Peninsula, Wales) |
| 25,000 | Earliest Ceramic Object ('Venus' of Dolní Věstonice, Moravia, Czech Republic) |
| 20,000 | Sea Level 125 metres below present at Last Glacial Maximum (LGM) |
| 19,000 | Abrupt end of Last Glacial Maximum in Europe (LGM started 26,500 BP) |
| 18,000 | Ceramic Pot Shards (Yuchanyan Cave, Hunan, China) |
| 17,500 | Spear Thrower (Combe Saunière, Dordogne) |
| 17,000 | Upper Paleolithic cave paintings (Altimira in Spain and Lascaux in France) |
| 14,000 | Early Natufian Culture in the Levant (pre-agricultural but semi-sedentary 'Villages') |
| 13,500 | Burials of Humans with Dogs (Bonn-Oberkassel and Early Natufian Ain Mallaha) |
| 13,000 | Jebel Sahaba, Quadan settled hunter-gatherer culture (evidence of warfare) |
| 13,000 | Early English Cave Art (Stag at Cresswell Crags and Reindeer on Gower Peninsula) |
| 12,900 | Younger Dryas stadial (abrupt cold dry interlude, mean global temperature drop 5°C) |
| 11,700 | Holocene Interglacial (warming period continues up to the present day) |
| 11,700 | Mureybet Tell, pre-pottery ('Neolithic Revolution' – Fertile Crescent, wild crops) |
| 11,000 | Göbekli Tepe Stone Carvings (settled hunter-gatherer Sanctuary, SE Anatolia) |
| 11,000 | Early Grave Goods (basalt beads in the Fertile Crescent) approximate date |
| 11,000 | Jericho first appeared (the world's first town, The Levant) |
| 10,800 | Earliest occupation of Amesbury in Wiltshire (dated from wild animal bones) |
| 10,500 | Domesticated Cattle (from wild Ox, *Aurochs*) in the area of NE Iraq & SE Turkey |
| 10,400 | Domesticated Wheat (from wild Emmer Wheat) in widespread use in the Levant |

| BCE | Pre-History *(BCE = years before current era)* |
|---|---|
| 8000 | World Population did not exceed 10 million before the invention of Agriculture |
| 7300 | South American Cave Art (hand stencils, Cueva de las Manos, Southern Argentina) |
| 7000 | Çatalhöyük, major Neolithic settlement, population 5,000+ (South Central Anatolia) |
| 6200 | Lake Agassiz drained through Hudson Bay (global sea level rise 0.8 to 2.8 m) |
| 6200 | Storegga Slide Tsunami (Doggerland submerged, England cut off from Continent) |
| 6000 | Neolithic Settlements in Egypt (perhaps due to increased desertification of the Sahara) |
| 5600 | Black Sea Deluge (Mediterranean broke through the Bosphorus) |
| 5500 | Lactase Persistence Allele (European mutation allowed adult milk digestion) est. date |

| | |
|---|---|
| 5000 | Samara Culture (Middle Volga Neolithic, included earliest horse burials) |
| 4400 | Earliest Nile Valley Farming Settlements (Badari Culture, Wheat, Barley, Cattle, Sheep) |
| 4200 | Burial of disabled dog with old spinal injury (Shamanaka, Cis-Baikal, Siberia) |
| 4100 | Uruk period in Sumer began (temple-centred cities and organised irrigation) |
| 4100 | Earliest known Winery (press and fermentation vats, Vayots Dzor, Armenia) |
| 4000 | Pottery and Farming first reached Southern Britain, approximate date |
| 3650 | West Kennet Long Barrow (Early Neolithic Chambered Tomb, Wiltshire) |
| 3500 | Harnessed Horses (evidenced by bit wear to teeth, Kazakhstan) |
| 3400 | Egytian Hieroglyphs (on ivory tags & seal impressions, Abydos) |
| 3350 | Kish Limestone Tablet (proto-cuneiform writing – Pictographic, Sumer) |
| 3300 | Nominal start of the Bronze Age in the Near East (Caucasus, Anatolia, Elam Susa) |
| 3250 | Earliest known Wheel (Ljubljana Marshes Wooden Wheel) ± 100 years |
| 3250 | 'Ötzi Ice Man' (natural mummy of man killed in the high Alps, found 1991) |
| 3200 | Ness of Brodgar (Neolithic sanctuary on Orkney) earliest date |
| 3180 | Skara Brae (Orkney 'Village') earliest date, occupied up to 2500 BCE approx. |
| 3150 | Narmer Palette evidenced unification of Upper & Lower Egypt |
| 3000 | Uruk Clay Tablets (impressed Proto-Cuneiform accounting information, Sumer) |
| 3000 | Stonehenge I, First Earthwork enclosure (ditch & bank with inner circle of posts) |
| 2850 | Avebury Henge and Stone Circles construction began (Wiltshire) |
| 2800 | Bell-Beaker Culture appeared (Tagus Valley Copper, Portugal) |
| 2650 | Stepped Pyramid of Saqquara (3rd dynasty Pharaoh Djoser, by his Vizier Imhotep) |
| 2600 | Mohenjo-Daro (major Indus Valley urban settlement, Mature Harrapan civilisation) |
| 2600 | 'Queen' Puabi buried in Ur with lyre, jewellery, gold headdress and 52 attendants |
| 2560 | Great Pyramid of Giza (Pharoah Khufu, son of 4th dynasty founder, Sneferu) |
| 2510 | Statue of Ka-aper, Priest of Saqquara (earliest life-sized wooden statue, Sycamore) |
| 2500 | Ebla Tablets referenced Canaanites, Ugarit and the production of beer (Syria) |
| 2500 | Stonehenge III Giant Sarsen Stones (Blue Stones perhaps 3000 BCE, Presili Hills) |
| 2475 | Bell-Beaker Folk arrived in Britain (Copper Age began, Cornish tin from 2000 BCE) |
| 2400 | Silbury Hill, Avebury, Wiltshire (size comparable to Pyramids, purpose unknown) |
| 2400 | 'Amesbury Archer' burial (Beaker vessels, copper & gold objects, Wiltshire) |
| 2375 | Earliest Pyramid Text (Pharoah Unas, Old Kingdom – referenced Ma'at) |
| 2350 | Early Dynastic period of Southern Mesopotamia ended, Earliest Ziggurats built |
| 2300 | Ness of Brodgar 'Cathedral' Structure Ten destroyed (masses of Cattle Bones) |
| 2270 | World's first Empire founded by Sargon of Akkad (conquered Sumerian City States) |
| 2225 | Chichester Burial with early Bronze Dagger, ± 75 years |

2200   West Kennet Long Barrow sealed off with Sarsen Stone

2100   Earliest known Chariot Burials (Sintashta Bronze Age Culture, Steppes SE of Urals)

2050   Code of Ur-Nammu, Sumerian (earliest extant law code – Justice based on Equity)

2049   'Seahenge' timber circle, dated by Oak trunks (bronze axes, Hunstanton, Norfolk)

2000   Great Ziggurat of Ur completed by King Shulgi son of Ur-Nammu, approximate date

2000   *Epic of Gilgamesh*, Earliest Sumerian Poems (from end of Third Dynasty of Ur)

2000   Coffin Texts (wealthy ordinary people) from First Intermediate Period, Egypt

2000   Callanish Stone Circle constructed (aligned with the Moon, Outer Hebrides)

1950   Minoan Crete Golden Age, through to 1450 BCE (Linear A Script not deciphered)

1900   Indus Valley cities collapsed and abandoned (rediscovered 1922 CE)

1850   Wadi el Hol graffiti, Egypt (earliest example of Proto-Canaanite alphabetic script)

1800   *Tale of Sinuhe*, earliest example (set around the death of Amenemhat I)

1780   Woolly Mammoth species extinct (last known population, Wrangel Island)

1730   Knossos Palace Complex destroyed by earthquake (again in 1570 BCE)

1772   Code of Hammurabi (Babylon, 282 Laws including #196: "An eye for an eye")

1700   Trundholm Sun Chariot (horse drawn with spoked wheels, bronze, West Zealand)

1700   Vedic Sanscrit Hymns (Aryan invasions of North West India) earliest date

1627   Thera Volcanic Explosion (est. 60 km$^3$, wall paintings and pottery preserved) ± 13 years

1620   Khyan, first named Hyksos King (Fifteenth Dynasty Egypt, died 1580 BCE)

1600   Mycenaean 'Golden Age' (Tholos tombs & golden grave goods) nominal start date

1600   Olmec rubber balls, stone heads, wooden figurine bog sacrifices (San Lorenzo)

1595   Hittites raided Babylon, peak of their power (Kassite allies re-took City 1531 BCE)

1551   Ahmose Pharoah, ejected Hyksos & reunited Egypt, died 1527 BCE, approximate dates

1550   'Book of the Dead', earliest example (start of the New Kingdom – Karnak Oracle)

1506   Thutmose I Pharaoh, Temple of Karnak at Thebes, conquest of Nubia (died 1493)

1500   Bronze Age Log Boats sunk in Flag Fen (Must Farm, Peterborough)

1500   Cult of Demeter, Greece (leading to Eleusinian Mysteries) approximate date

1473   Hatshepsut, female Pharoah in minority of Thutmose III (died 1458 BCE)

1375   Minoan Palaces final destruction (control passed to Mycenaeans, Linear B)

1353   Akhenaten Pharoah (Monotheistic Aten worship, Amarna Letters, died 1336 BCE)

1332   Tutankhamun died (rich grave goods, tomb found 1922 CE by Howard Carter)

1305   Uluburun Bronze Age Shipwreck (cargo bound for Greece from the Levant)

1300   The Exodus, Moses and the Ten Commandments, supposed date

1274   Battle of Kadesh (Hittites, chariot battle, Ramesses II tomb wall, Treaty 1259 BCE)

1200   Destruction of Pylos & Mycenae (perhaps by 'Sea Peoples') approximate date

| | |
|---|---|
| 1200 | Earliest known Israelite settlements in Central Hill Country (*per* Isaac Finkelstein) |
| 1200 | Rigveda final form, Sanscrit Brahmanic oral text (Vedic culture reached Ganges Plain) |
| 1200 | *Epic of Gilgamesh*, Akkadian version (from Library of Ashurbanipal in Nineveh) |
| 1184 | Fall of Troy (Level VIIa) possible date (but could be earlier Troy VIf/g, *ca.* 1400 BCE) |
| 1177 | Battle of the [Nile] Delta, Ramesses III defeated the 'Sea Peoples' (Temple Mural) |
| 1150 | Late Bronze Age Collapse (Mycenaeans, Hittites and New Kingdom Egypt) end date |
| 1050 | Earliest Phoenician Alphabetic inscription (Sarcophagus of King Ahirom of Byblos) |

## BCE            Early Recorded History  *(BCE = years before current era)*

| | |
|---|---|
| 970 | King David died (succeeded by Solomon, his son by Bathsheba) |
| 950 | Tel Zayit inscription (Paleo-Hebrew alphabet, 22 consonants) approximate date |
| 928 | Division of Northern Kingdom of Israel from Judah to the South |
| 900 | Olmec giant clay Pyramid and urban plan (La Venta, Mexico) |
| 840 | Mesha Stele (corroborates 2 Kings 3 defeat of Israel by Moab) |
| 800 | Deir Alla Inscription (alphabetic, polytheistic, prophesy of the Moabite, Bala'am) |
| 776 | First Games at Olympia (continued 4-yearly, banned by Theodosius 393 CE) |
| 730 | Dipylon Pottery Jug Inscription (early use of Greek alphabet, ± 20 yrs, Athens) |
| 732 | First Assyrian invasion, 'Lost Tribes' of Israel disappeared into captivity and oblivion |
| 716 | Hezekiah's reign started (died 687 BCE) compilations of JE then P Bible sources |
| 701 | Siege and defeat of Lachish by Assyrians (but Sennacherib failed to take Egypt) |
| 700 | *The Iliad*, written down by or for Homer from oral recitation, approximate date |
| 700 | *Works and Days* and *Theogony*, written by Hesiod, approximate date |
| 667 | Assyrians conquered Egypt (peak of Assyrian Empire under Ashurbanipal) |
| 650 | Carthage became independent Phoenician City-state (founded by Tyre 814 BCE) |
| 620 | Disintegration of Assyrian Empire (following death of Ashurbanipal 627 BCE) |
| 609 | King Josiah died in Battle of Megiddo against Egyptians (Temple renovation 622 BCE) |
| 606 | Pittacus of Mytilene (Lesbos) defeated Phrynon (died 568 BCE) 'The Golden Rule' |
| 605 | Babylonians and Medes finally defeat Assyrians and Egyptians at Carchemish |
| 587 | Fall of Jerusalem, third and final deportation to Babylon (The first was 597 BCE) |
| 583 | Zarathustra died at Balkh (Persian founder of Zoroastrianism) approximate date |
| 574 | Solon's Reforms (basis for republic, Athens) possible date (died 558 BCE) |
| 570 | Sappho of Lesbos died (romantic poet) approximate date |
| 550 | Croesus King of Lydia minted Gold Coins (defeated by Persians 546 BCE) |
| 546 | Thales died, first pre-Socratic Philosopher (born 624 BCE) 'The Golden Rule' |

540   New Temple of Artemis completed (Ephesus, Lydia) approximate date

538   Cyrus the Great overran Babylon (Jews allowed to return to Judah, but many stayed)

528   The Buddha attained enlightenment, died 483 BCE aged 80 (dates controversial)

525   Persians conquered Egypt (Elephantine papyri, Jewish community 495 to 399 BCE)

508   Athenian Democracy (Cleisthenes, Popular Assembly followed revolution)

500   *Tao Te Ching*, thought to have been composed by Laozi, uncertain date

495   Pythagoras of Samos died at Metapontium (taught by Delphic Priestess Themistoclea)

490   Battle of Marathon (Athenian army defeated the first Persian invasion)

480   Battle of Thermopylae, then Salamis (Persian fleet destroyed) Delian League 478 BCE

479   Confucius died (born 551 BCE, correct behaviour, respect and 'The Golden Rule')

475   Heraclitus of Epesus died, pre-Socratic Philosopher (born 535 BCE)

474   *Charioteer of Delphi*, bronze statue in the Sanctuary of Apollo

472   *The Persians*, Aeschylus, the earliest extant Tragedy (presented by Pericles)

460   Parmenides of Elea died (only fragments of his poem 'On Nature' survive)

447   Parthenon construction initiated by Pericles (completed 438 BCE)

445   Nehemiah (cup-bearer to Artaxerxes) Persian Governor of Judah (Ezra, High Priest)

429   Pericles died in Athenian Plague (Peloponnesian war began 431 BCE)

427   'Mytilenian Debate' (Athenian Assembly called off reprisals against city-state)

425   Herodotus died (author of *The Histories*, credited as the first historian)

415   *The Trojan Women*, Euripides (against background of the sack of Melos by Athens)

413   Sparta won Battle of Syracuse (Sicilian Expedition ended in disaster for Athens)

408   *Orestes*, last Tragedy of Euripides performed in his lifetime (died 406 BCE)

404   Athens finally surrendered to Sparta (Peloponnesian War recorded by Thucydides)

399   Socrates' trial and execution (born 469 BCE) recounted by Plato and Xenophon

380   Plato's *Republic*, *Phaedrus* and *Symposium* written around this date

377   Hippocrates died (credited as the first to attribute disease to natural causes)

364   Praxiteles' Sculptures including first life-size female nude (dated by Pliny the Elder)

348   Death of Plato (born 428 BCE, his Academy endured until Mithridatic War 88 BCE)

336   Philip II of Macedon assassinated (succeeded by his son Alexander, then aged 20)

332   Alexander founded Alexandria (after defeating Persian Empire under Darius III)

323   Alexander the Great died aged 32 in Babylon (Hephaestion had died 324 BCE)

322   Aristotle died (born 384 BCE, Plato Academy to age 37, tutor to Alexander)

322   Chandragupta established Maurya Empire (took Magadha and defeated the Greeks)

300   *Elements*, Euclid of Alexandria (geometry and mathematics) approximate date

283   Ptolemy II King of Egypt (built Library of Alexandria, translated Torah into Greek)

| | |
|---|---|
| 240 | Vertical Waterwheel Powered Mill (Alexandria), probable date |
| 232 | Ashoka died aged 72, Buddhist Emperor (33 Edicts, including to protect Animals) |
| 212 | Archimedes of Syracuse died aged 75 (probably designed Antikythera Mechanism) |
| 210 | Qin Shi Huang died (First Emperor of China, buried with Terracotta Army) |
| 202 | Han Dynasty founded by rebel Liu Bang, Emperor Gao (Han period ended 202 CE) |
| 200 | Iron Age Bog Burials of Human Sacrifices (Ireland, Netherlands, Denmark) |
| 168 | Third Macedonian War ended (Roman General Paullus defeated Perseus of Macedon) |
| 146 | Carthage destroyed by Rome and in the same year the Battle of Corinth |
| 125 | Zhāng Qiān returned to Han China from expedition along what became the Silk Road |
| 86 | Athens defeated following siege by General Sulla, First Mithridatic War ended |
| 71 | Spartacus finally defeated (Third Servile War ended, 6,000 survivors crucified) |
| 55 | Julius Caesar landed in Britain (Rubicon 49, assassinated, Ides of March 44 BCE) |
| 31 | Octavian (Augustus) defeated Mark Antony and Cleopatra at Actium |
| 6 | Jesus of Nazareth was born (probable date – Herod the Great died 4 BCE) |

CE       The First Millenium  *(CE = years of the current era)*

| | |
|---|---|
| 1 | World Population estimate 230 million (Roman Empire perhaps 14 million) |
| 9 | Roman army defeated by Arminius (Teutoburg Forest, Varus committed suicide) |
| 10 | Hillel the Elder died in Jerusalem (Jesus probably aged 15) 'The Golden Rule' |
| 33 | Jesus Christ crucified (possible date) *"Love ... thy neighbour as thyself"* |
| 43 | Romans invaded Britain (Emperor Claudius established Colchester as capital) |
| 70 | Second Temple destroyed and dispersion of the Jews began |
| 79 | Eruption of Vesuvius buried Pompeii and Herculaneum (Pliny the Elder died) |
| 105 | Paper making from pulp composition invented (Cai Lun at Eastern Han Court) |
| 180 | Emperor Marcus Aurelius died at Vindobona in Austria (Stoic *Meditations*) |
| 210 | Galen died (Greek physician to Roman Emperors, applied medical investigations) |
| 313 | Edict of Milan (Constantine I and Licinius agreed religious tolerance) |
| 313 | Sogdian Ancient Silk Road Letters (Dunhuang, found by Aurel Stein, 1907) |
| 325 | Council of Nicaea (convened to deal with Arian Controversy re the Trinity) |
| 330 | Constantinople founded by Emperor Constantine (died 337) |
| 395 | Augustine, Bishop of Hippo ('Original Sin', died 430 during Vandals' siege, Saint 1298) |
| 410 | Alaric, Visigoths sacked Rome, Roman rule ended in Britain (Ravenna capital 402) |
| 415 | Hypatia of Alexandria (head of Neoplatonist school) murdered by Christian mob |
| 449 | Anglo-Saxon invasions started (Hengist & Horsa landed in Kent as mercenaries) |

455   Geiseric the Vandal sacked Rome (Arian Christianity, captured Carthage in 439)

500   Ostragoth King Theodoric built Arian Baptistry (Ravenna, Mosaics)

529   Neoplatonist Academy closed by Justinian (members went to Sassanid Ctesiphon)

529   Benedict of Nursia founded Monte Cassino on a Pagan site (Rule of Saint Benedict)

536   Extreme Cold Weather Event and global famine (space impact or volcanic dust)

537   Basilica of Hagia Sophia inaugurated by Emperor Justinian (Constantinople)

541   Plague of Justinian (Constantinople mortality 40%) Periodic recurrences until 750

552   Buddhism introduced to Japan from China *via* Korea (official date per Ihon Shoki)

563   Christian monastery on the Island of Iona founded by Columba (from Ireland)

574   Áedán mac Gabráin King of Dál Riata (Dalriada) Contemporary of St. Columba

597   Augustine of Canterbury converted Kingdom of Kent to Christianity (Pope Gregory)

618   Tang Dynasty founded (capital Chang'an became the most populous city in the world)

628   Introduction of digit zero, (Brahmagupta, Indian mathematician, died 670)

630   Sutton Hoo Ship Burial (possibly of King Rædwald, died 626)

632   The Prophet Muhammad died (Hijra to Medina 622, conquered Mecca 630)

664   Synod of Whitby (adopted Roman Easter) Cuthbert Prior of Lindisfarne (died 687)

680   Battle of Karbala (defeat of the Prophet's grandson Ali by Umayyad Caliph)

697   First Doge of Venice (Paolo Lucio Anafesto) later the 'Serene Republic'

698   *Lindisfarne Gospels* (produced in Latin by Eadfrith, Bishop of Lindisfarne)

698   St. Cuthbert Gospel (of St. John) put in the Saint's coffin as he was re-interred

711   Muslim Berbers captured Cordoba (Al-Andalus – all Iberia except Asturias by 720)

731   *History of the English People*, Venerable Bede, completed (died 735)

749   John of Damascus died (Defender of Holy Images, born Yuhanna ibn Mansur)

754   Council of Hieria, Iconoclasm confirmed by Constantine V (ended 787)

755   Golden Age of Tang Dynasty ended by An Lushan Rebellion (Emperor Xuanzong)

756   Abd al-Rahman Emir of Córdoba (Umayyads having fled Abbasid Damascus in 750)

762   Baghdad founded by Al Mansur as Abbasid capital (started 'Golden Age of Islam')

776   Hindu–Arabic Number System adopted by Abbasid Caliph Al-Mansur

777   Abou Ben Adhem died, ascetic Sufi Saint (subject of Leigh Hunt Poem 1838)

782   Massacre of Verden (Charlemagne executed 4,500 Saxon captives)

787   Second Council of Nicaea (ended Byzantine Iconoclasm)

793   Vikings first raided Lindisfarne (Monks left 875 carrying the bones of St. Cuthbert)

800   Charlemagne crowned Emperor in Rome (King of the Franks 768, died 814)

800   *Beowulf* (epic poem in Old English, set in Scandinavia) proposed date

834   Oseberg Ship Burial of high-status woman (Tønsberg, SSW of modern day Oslo)

| | |
|---|---|
| 841 | Viking Settlement of Dublin (European Slave Market, formerly Christian Dubhlinn) |
| 866 | Vikings captured York (Jorvik) led by Ivar the Boneless |
| 878 | Battle of Edington, Alfred defeated Guthrum (Danelaw agreed by treaty 886) |
| 882 | Kievan Rus' established by Prince Oleg (Christian from 990, Max. extent 1050) |
| 899 | King Alfred died (two or three years after his final victory over the Danes) |
| 911 | Treaty of Saint-Clair-sur-Epte (Norse Rollo became Robert Duke of Normandy) |
| 910 | Cluny Abbey founded (Benedictine leader of Western monasticism) |
| 927 | Æthelston (Alfred's grandson) acknowledged Overlord of all the English Kings |
| 929 | Abd al-Rahman III declared Caliph in Córdoba (Umayyad, Al-Andalus) |
| 930 | *Althing* established as Icelandic Parliament (first Viking Settlement 874) |
| 970 | Old English Gloss of *Lindisfarne Gospels* by Aldred (Chester-le-Street, Durham) |

| | |
|---|---|
| 1000 | World Population estimate 300 million |
| 1000 | Norse Settlement on Newfoundland (L'Anse aux Meadows) Lief Ericson, Vinland |
| 1000 | *Ælfric's Colloquy* in Latin and Anglo-Saxon incl. 'The Saxon Ploughman's Lament' |
| 1013 | Medina Azahara (Córdoba) sacked by mob (Córdoba fell to Christians in 1236) |
| 1035 | Cnut died (King of England from 1016, Denmark 1018 and Norway 1030) |
| 1066 | Norman Conquest (Battle of Hastings, hard after Battle of Stamford Bridge) |
| 1086 | 'Doomsday Book' completed (comprehensive assessment of English land holdings) |
| 1093 | Winchester Cathedral consecrated (Norman, replaced Anglo-Saxon founded 642) |
| 1099 | First Crusade captured Jerusalem (last Christians driven out of Holy Land 1303) |
| 1111 | al–Ghazālī died, Persian Moslem theologian (approved Sufism, Born 1058) |
| 1131 | Omar Khayyam died (Persian polymath, Rubáiyát by Edward Fitzgerald pub. 1859) |
| 1167 | 'Assize of Clarendon' began change to jury based law (Henry II, after English civil war) |
| 1167 | Oxford University founded on or before this date (Cambridge founded 1209) |
| 1204 | Maimonides died in Egypt (Philosopher & Physician, born in Córdoba 1135) |
| 1209 | Francis of Assisi began to preach (died 1226, Saint Francis 1228) |
| 1215 | 'Magna Carta' forced on King John at Runnymede (sealed not signed 15th June) |
| 1227 | Genghis Khan died, 'Great Khan' (founded Mongol Empire 1206, born *ca.* 1162) |
| 1240 | Chartres Stained Glass Windows completed (all but 3 earlier ones ruined in 1195) |
| 1242 | Ibn al-Nafis described blood circulation (referenced Avicenna's work from 1025) |
| 1252 | Alfonso X Crowned King of Castille (founded School of Translators of Toledo) |
| 1258 | Baghdad sacked by Genghis Khan's grandson (ended 'Golden Age of Islam') |

| | |
|---|---|
| 1260 | Mongols defeated by Mamluk Sultanate at Ain Jalut (limited the Mongol Empire) |
| 1265 | de Montfort's Parliament (followed by Edward I model Parliament 1295) |
| 1266 | Nicolò & Maffeo Polo reached Court of Kublai Khan (Marco voyaged 1271–1295) |
| 1273 | Rumi died (Jalāl ad-Dīn Muhammad Balkhī) Persian Sufi Poetry & Dance |
| 1286 | Spectacles invented in Italy (dated by Giordano da Pisa 1306 sermon) |
| 1290 | Jews expelled from England (Gregory of Huntingdon saved Hebrew Texts) |
| 1291 | Swiss Confederation established (three Cantons, expanded to eight by 1353) |

## CE         The Renaissance   *(CE = years of the current era)*

| | |
|---|---|
| 1305 | Giotto completed the frescoes of the Arena Chapel (Padua) |
| 1315 | Famine in Europe from extreme cold wet weather (effects extended for a decade) |
| 1321 | Dante Alighieri died in exile in Ravenna (author of *The Divine Comedy*) |
| 1348 | Black Death (Plague) estimated World population of 450M reduced to 350M |
| 1356 | 'Diet of the Hansa' at Lübeck established Hanseatic League (last meeting 1669) |
| 1369 | Ibn Battuta died, Berber scholar (travelled Islamic world from Timbucktu to Beijing) |
| 1380 | John Wycliffe, first English Bible translation (Lollards, precursor to Reformation) |
| 1381 | Peasants Revolt, Poll Tax in Kent, Wat Tyler killed (beginning of the end of Serfdom) |
| 1382 | Moscow fell to the Golden Horde (Tartars killed 24,000 Muscovites) |
| 1382 | Winchester College founded by William of Wykeham, Bishop of Winchester |
| 1390 | Last English Wolf killed (Humphrey Head, Cumbria) |
| 1398 | Tamerlane (Timur the Lame) sacked Delhi (executed 100,000 Hindu captives) |
| 1400 | Geoffrey Chaucer died (*Canterbury Tales*, first printed by Caxton 1478) |
| 1405 | Chinese eunuch Admiral Zheng He led the first of seven voyages to the Indian Ocean |
| 1434 | Cosimo de' Medici returned to Florence (Banking dynasty lasted three centuries) |
| 1449 | Letters Patent under the King's Great Seal to John of Utynam for Stained Glass |
| 1450 | Western Norse Settlement on Greenland abandoned (Eastern abandoned 1350) |
| 1453 | Constantinople fell to Ottoman Turks (ended Roman Empire, early use of Cannon) |
| 1454 | Gutenberg Bible printed in Latin (first use of moveable type, Mainz, Germany) |
| 1475 | Treaty of Picquigny formally ended Hundred Years War (English left France) |
| 1476 | Printing Press established in Westminster by William Caxton (derivation Cologne) |
| 1482 | *Primavera* by Botticelli (painting so named by Giorgio Vasari in 1550) |
| 1492 | Alhambra Palace & Islamic Emirate of Granada surrendered to Catholic Monarchs |
| 1492 | Jews and Muslims expelled from Spain at four months notice (from Portugal 1497) |
| 1492 | Christopher Columbus first voyaged to the Americas |

1494   Double Entry Book-keeping method published (Luca Pacioli, Venice, Died 1517)

1495   Albrecht Dürer opened his workshop in Nuremburg (prints from intricate woodcuts)

1497   Books & paintings burned in Florence by Dominican priest Savonarola

1498   *The Last Supper* completed by Leonardo da Vinci (Mural, Milan)

1500   *World Population estimate 500M (0.5 billion)*

1504   *David* completed by Michelangelo (Carrara Marble Statue, Florence)

1517   Selim I, First Ottoman Caliph (defeated Mamluk Sultanate of Egypt the same year)

1517   Martin Luther nailed 95 'Theses of Contention' to Wittenberg Church Door

1521   Spanish Conquistadores destroyed Aztecs (Cortés, then Pizzaro destroyed Incas, 1532)

1522   Circumnavigation of the world (Elcano with 17 survivors of Magellan's expedition)

1525   English New Testament first printed (William Tyndale)

1526   Babur founded Mughal Empire (descendant of Timur and of Genghis Khan, died 1530)

1532   *The Prince*, Nicolò Machiavelli, published posthumously (died 1527)

1536   *Institutes of the Christian Religion*, John Calvin (died 1564)

1536   Dissolution of Monasteries by Henry VIII (2nd Act of Suppression was 1539)

1543   *Revolution of the Celestial Spheres*, Nicolaus Copernicus (died the same year)

1547   Ivan 'The Terrible' first Tsar of all the Russians (Oprichnina secret police 1565 to 1572)

1580   *Essays*, Michel de Montaigne, first part published (written 1570 to 1592)

1588   Spanish Armada (failed attempt to invade England and depose Queen Elizabeth)

1598   Edict of Nantes (freedom of religion for Huguenots in France, Revoked 1685)

1600   Huaynaputina volcanic eruption (Peru), Famine killed one third of Russian population

1600   Giordano Bruno burned for Heresy (proposed the Sun as Star and infinite Worlds)

1600   East India Company (company rule in India from 1757, British Raj from 1858)

1602   Dutch East India Company (Chartered monopoly in Asian trade) Dissolved 1800

1605   *Don Quixote*, Part One published by Miguel de Cervantes (part two 1615)

1607   *L'Orfeo* by Claudio Monteverdi premiered in Mantua (*Vespers* 1610)

1607   Virginia, first of the thirteen British Colonies on North American East Coast

1610   *Sidereus Nuncius* published, Galileo Galilei (Telescopic Astronomy, died 1642)

1611   King James Bible (Old Testament translated from Hebrew and New from Greek)

1616   William Shakespeare died (born 1564, *Hamlet* 1601, First Folio published 1623)

1624   Rembrandt van Rijn opened his first studio in Leiden, Holland (died 1669)

1627   Wild Ox (*Aurochs*) became extinct (last known animal died in Poland)

1630   *Miserere* composed by Allegri for Sistine Chapel (memorised by Mozart 1770)

1632   Taj Mahal begun by Mughal Emperor Shah Jahan (mausoleum for his 3rd wife)

1637   *"Cogito ergo sum"*, Réné Descartes (philosopher & mathematician, died 1650)

1646  *The Great Art of Light and Shadow*, Kircher, described *Camera Obscura*

1648  'Peace of Westphalia' ended 30 years war (recognised the Protestant Dutch Republic)

1649  King Charles I executed during the English Revolution (Oliver Cromwell, died 1658)

1652  Dutch colony founded at Cape of Good Hope (Dutch East India Co. re-supply base)

1653  *The Compleat Angler*, Isaak Walton, first published

1656  Jewish Return to England allowed informally by Oliver Cromwell


CE                     The Age of Enlightenment  *(CE = years of the current era)*

1660  Restoration of the Monarchy (Charles II) proclaimed by Parliament

1660  Royal Society founded "to promote excellence in science …"

1665  *Girl with a Pearl Earring*, Vermeer (who perhaps used *Camera Obscura*)

1665  *Micrographia*, Robert Hooke, published (he later introduced the term 'Cell')

1665  Great Plague of London (last Bubonic epidemic) ended by Great Fire of 1666

1677  Spinoza died (rationalist, *The Ethics* published posthumously, Amsterdam)

1682  Bideford Witch Trial, Exeter, Devon (Salem Witch Trials, Massachusetts, 1692)

1683  Battle of Vienna (apogee of Ottoman invasion of Europe)

1683  Frost Fair, River Thames froze ('Maunder Minimum', froze again 1788 and 1814)

1685  'Edict of Fontainebleau' against Protestants in France (revoked 1598 Edict of Nantes)

1687  *Principia Mathematica*, Isaac Newton, published (gravity supposed 1666)

1688  Lloyd's Coffee House, started the London Insurance Market

1689  'Bill of Rights' enacted in England ('Petition of Right' 1628, 'Habeas Corpus' 1679)

1693  John Locke argued against cruelty to animals (contradicting Descartes and Kant)

1694  Bank of England Banknotes first issued (handwritten)

1694  Matsuo Basho died in Osaka, Haiku poet (born in Ueno 1644)

1698  'Merchant Venturers' of Bristol, broke into West African–West Indes Slave Trade

1700  Pianoforte first appeared, made by Cristofori of Padua for Ferdinando de' Medici

1701  Horse Drawn Seed Drill invented by Jethro Tull (died 1741)

1702  *The Daily Courant*, the first English daily newspaper published (*The Times*, 1788)

1703  Tsar Peter the Great founded St. Petersburg (tens of thousands of serfs died)

1704  'Darley Arabian' imported from Syria (foundation sire to 80% of Thoroughbreds)

1707  'Act of Union' between England and Scotland (United Monarchy from 1603)

1713  Peace of Utrecht ended War of Spanish Succession (checked French influence)

1720  'South Sea Bubble', London Stock Exchange crash (and 'Mississippi Bubble', Paris)

1722  Europeans reached Easter Island (population had crashed from 15K to 3K in a decade)

1730   'Turnip' Townshend's Four-Field Crop Rotation (on leaving politics, Died 1738)

1735   Witchcraft Act effectively ended Witch Trials in England

1737   Antonio Stradivari died in Cremona (born 1644, Violin 'Golden Age' 1700 to 1725)

1746   Culloden Moor ended the 1745 Jacobite Rebellion (the last battle on British soil)

1748   Pension Fund based on actuarial probability (Webster & Wallace, Scottish Widows)

1749   *Mass in B Minor* by Johann Sebastian Bach completed (died the following year)

1750   *Messiah* by Handel, first annual charity performance for Coram's Foundling Hospital

1755   *A Dictionary of the English Language*, compiled by Samuel Johnson (died 1784)

1759   *Candide*, Voltaire, published (François-Marie Arouet 'Voltaire' died 1778)


CE        The Industrial Revolution *(CE = years of the current era)*

1761   Marine Chronometer, Atlantic trial (John Harrison, inventor, then aged 68)

1761   Bridgewater Canal opened (coal to Manchester, James Brindley engineer)

1762   *Whistlejacket*, painted by George Stubbs (grandsire: Godolphin Arabian b.1724)

1764   Blenheim Palace Gardens landscaped by 'Capability' Brown (died 1783)

1771   *Encyclopædia Britannica*, first three-volume edition published in Edinburgh

1771   Water-powered Spinning Frame ('First Factory' Cromford Mill, Richard Arkwright)

1772   Lord Mansfield's Judgement first emancipated a slave in England

1773   'Inclosure Act' introduced procedures for landowners to enclose Common Land

1774   "*Speech to the Electors at Bristol ...*" Edmund Burke, representative government

1775   Water Closet (Toilet) with S-Trap in drain (flush tank with float valve 1778)

1775   Steam Engine built by James Watt (improved on Newcomen Engine of 1712)

1776   *The Wealth of Nations*, Adam Smith, first published (Industrial Economics)

1776   Declaration of Independence from Britain by thirteen American Colonies

1780   Lincoln Longwool then Leicester Sheep breeding by Robert Bakewell (died 1795)

1781   Iron Bridge over the Severn opened (built of Cast Iron by Abraham Darby III)

1783   Laki volcanic eruption (Iceland) killed up to 6M including 1M in France

1783   'Peace of Paris' ended American Revolutionary War (USA, Britain, France & Spain)

1784   Mail Coaches established with a record run from Bristol to London in 16 hours

1785   Vertical Power Loom patented by Edmund Cartwright

1786   Moses Mendelssohn died in Berlin age 57 (German-Jewish philosopher, the Hashkalah)

1787   *Don Giovanni,* Wolfgang Amadeus Mozart, premiered in Prague (Mozart died 1791)

1789   *Declaration of the Rights of Man* and Storming of the Bastille (14 July)

1789   *The Natural History of Selborne*, Gilbert White, first published (died 1793)

1793   British embassy to China failed (Macartney – Emperor Qianlong)

1794   *Age of Reason*, Thomas Paine, published (followed *Rights of Man*, 1791)

1796   Smallpox Vaccination pioneered by Dr. Edward Jenner (declared eradicated 1979)

1798   *An Essay on the Principle of Population*, Thomas Robert Malthus, published


CE          The Romantic Era  *(CE = years of the current era)*

1800   Anthropocene (humans determine Earth's ecosystems) atmospheric $CO_2$ 280 ppm

1800   World Population reached 1 billion, London 1M, England 17% Urban

1801   Jacquard Punch-card driven Loom demonstrated (Cards later inspired Babbage)

1805   Battle of Trafalgar established British Naval supremacy (death of Lord Nelson)

1810   'Comet', Beef Shorthorn Bull, bred by Charles Colling, sold for 1,000 Guineas

1811   *Sense and Sensibility*, Jane Austen, published (romantic novel) died 1817

1812   Napoleon's retreat from Moscow (Napoleonic War ended at Waterloo, 1815)

1815   Mount Tambora eruption (est. 160 km$^3$, Indonesia) famine in Europe, 1816

1817   *Matrikellisten* In Lower Franconia (Wurzburg) after 'Bavarian Jews Edict', 1813

1818   Prussian Customs Union (*Zollverein*) basis for expanded German Confederation

1819   *Ode to the West Wind*, Shelley (died 1822 age 29 – Keats died 1821, of TB age 25)

1821   *The Haywain*, by John Constable shown at the Royal Academy (died 1837)

1822   'Act to Prevent the Cruel and Improper Treatment of Cattle' (Dick Martin MP)

1824   Lord Byron (George Gordon) died of Malaria at Missolonghi aged 36

1824   *Ninth Symphony* by Ludvig van Beethoven premiere in Vienna (died 1827)

1825   Stockton to Darlington Railway (George Stephenson, first locomotive 1814)

1827   Battle of Navarino (Pylos) Ottoman Navy defeated (Greek independence 1832)

1828   Electric Motor demonstrated by Ányos Jedlik (DC, stator, rotor & commutator)

1829   Metropolitan Police formed by Robert Peel (Home Secretary)

1833   Slavery Abolition Act covering the British Empire (enforced by Royal Navy)

1832   *Faust*, Johann Wolfgang von Goethe (died the same year, published posthumously)

1837   'Analytical Engine' (general purpose computer) described by Babbage (Ada Lovelace)

1837   Alexander Pushkin died following a duel (*Eugene Onegin* published 1831)

1838   London to Birmingham railway opened (London terminus at Euston)

1839   *The Fighting Temeraire*, exhibited by J M W Turner (died 1851)

1840   Uniform Penny Post established throughout the UK (London Penny Post 1680)

1841   Photographic Positive-Negative, William Henry Fox Talbot (Daguerreotype 1837)

1842   'Mines and Collieries Act', Lord Ashley later Lord Shaftesbury (died 1885)

| | |
|---|---|
| 1842 | First Afghan War ended (Khyber Pass, Surgeon Brydon sole survivor of 16,000 men) |
| 1845 | 'SS Great Britain' entered service Bristol–New York (Isambard Kingdom Brunel) |
| 1845 | Potato Blight, Irish Famine exacerbated by Corn Laws, *ca.* 1M died and 2M emigrated |
| 1846 | General Anæsthetic demonstrated in public (William Morton, Massachusetts) |
| 1847 | *Jane Eyre* (Charlotte) and *Wuthering Heights* (Emily) by Brontë Sisters, published |
| 1848 | 'Year of Revolutions' France, Germany, Italy, Poland & the Austrian Empire; all failed |
| 1848 | 'Communist Manifesto' published in German by Karl Marx and Friedrich Engels |
| 1848 | Pre-Raphaelite Brotherhood formed (Holman Hunt, Millais & Rossetti) |
| 1849 | Fyodor Dostoyevsky death sentence for sedition commuted by Tsar Nicholas I |

## CE        The Age of Empires *(CE = years of the current era)*

| | |
|---|---|
| 1851 | *Moby Dick*, Herman Melville (a Sperm Whale sank the 'Essex' of Nantucket, 1820) |
| 1852 | *Uncle Tom's Cabin*, Harriet Stowe published (US anti slavery novel) |
| 1853 | *La Traviata*, Guiseppe Verdi, premiere La Fenice, Vienna (London & New York, 1856) |
| 1853 | Vapour-compression Refrigerator, Patent filed by Alexander C. Twining |
| 1854 | *Walden*, Henry David Thoreau, published (mentor R W Emerson) died 1862, TB age 44 |
| 1854 | London Broad Street Cholera Outbreak (John Snow traced source to sewage) |
| 1855 | 'Limited Liability Act' protected 25+ Joint Stock Co. members (7+ from Co's Act 1862) |
| 1855 | Fall of Sevastopol, turned Crimea War (telegraph, photos, Florence Nightingale) |
| 1855 | David Livingstone reached Victoria Falls (first European to do so) died 1873 |
| 1857 | 'Indian Rebellion', Sepoys challenged East India Company (led to British Raj, 1858) |
| 1859 | First Commercial Oil Well drilled (Drake Well, Titusville, Pennsylvania) |
| 1859 | *On the Origin of Species*, Charles Darwin published (born 1809, died 1882) |
| 1861 | Serfdom abolished in Russia (on unfavourable terms) |
| 1862 | 'First Homestead Act' (any US adult incl. freed Slaves could apply for 160 acres) |
| 1862 | East Coast Railway, London to Edinburgh 10½ hours (Flying Scotsman 1928) |
| 1863 | Impressionists, 'Salon des Refusées' (subsequent exhibitions 1875 and 1886) |
| 1863 | First Underground Railway opened (London Paddington to Faringdon) |
| 1863 | English Football Association formed (universally accepted game rules) |
| 1865 | *Tristan und Isolde*, Richard Wagner, Premiere in Munich |
| 1865 | *Experiments on Plant Hybridization*, Gregor Mendel described inheritance |
| 1865 | American Civil War ended (started 1861) but Abraham Lincoln assassinated |
| 1866 | First workable Transatlantic Cable (telegraph used Morse Code) |
| 1867 | *Das Kapital*, Karl Marx, published (critique of political economy) |

1869   Transcontinental Railway (from Omaha, Nebraska to Sacramento, California)

1869   Suez Canal opened (available to all nations in peace and war by 1888 Treaty)

1869   Girton College Cambridge (first residential university college for women)

1869   *War and Peace*, Leo Tolstoy (*Anna Karenina*, 1877) died 1910

1871   Fall of Paris following siege by German forces, ended Franco-Prussian War

1871   Unification of Germany (Bismarck, Chancellor) Wilhelm of Prussia declared Emperor

1871   Rome became capital of the Kingdom of Italy (Victor Emanuel II)

1872   Yellowstone, the first National Park designated in the United States

1876   Rubber Tree seed taken to Kew and thence to India and Malaysia (Henry Wickham)

1876   Telephone Patent granted to Alexander Graham Bell

1878   Incandescent Bulb Patent granted to Thomas Edison (second Patent 1879)

1880   *The Brothers Karamazov*, Fyodor Dostoyevsky, published (died 1881)

1880   'Education Act', compulsory schooling for ages 5–10, ended child labour

1881   The first Public Electricity Supply (Godalming, Surrey, England)

1883   Krakatoa Volcanic Explosion (Indonesia, killed 40,000+, heard 3,000 miles away)

1883   North American Bison reduced to 300 animals (from est. 30M – 60M in 1500)

1884   Berlin Conference (colonisation of Africa agreed by European Powers – Bismark)

1884   Steam Turbine driven Generator invented by Charles Parsons (7.5 kW)

1884   Petrol Engine invented by Edward Butler (incl. spark plug, magneto and carburettor)

1884   *Adventures of Huckleberry Finn*, Mark Twain, published in England (USA 1885)

1885   *Thus Spake Zarathustra*, Friedrich Nietzsche, final part published (died 1900)

1887   Columbia Records began operations (phonograph cylinders, flat discs from 1901)

1888   Pneumatic Tyre developed by John Boyd Dunlop (Vulcanisation by Goodyear, 1844)

1890   *News from Nowhere*, William Morris, designer (socialist utopia) died 1896

1890   *Hedda Gabler*, Henrik Ibsen (Norwegian realist playwright, died 1906)

1890   *Wheatfield* series by Vincent van Gogh at Arles (died the same year aged 37)

1893   Electric Waves (*i.e.* Electromagnetism) by Heinrich Herz (died 1894 aged 36)

1893   *9th Symphony* by Dvorak premiere at Carnegie Hall, which opened in 1891

1893   *6th Symphony* by Tchaikovsky premiere in St Petersburg (he died 9 days later)

1893   *The Scream* in both Paint and Pastel versions by Edvard Munch (died 1944)

1894   'Drefus Affair', of Jewish descent, falsely convicted of treason (exonerated 1906)

1895   Underwood Typewriter, four rows of keys and paper surface visible (QWERTY 1873)

1895   X-rays discovered by Wilhelm Röntgen and subsequently their medical use

1897   Vienna Secession formed ('Beethoven Frieze' by Gustav Klimt, 1902)

1898   *The Seagull,* Anton Chekhov, opened at Moscow Arts Theatre (died 1904)

1900  Eastman Kodak 'Box Brownie' (inexpensive 'snapshot' camera using rollfilm)

1900  Last wild Passenger Pigeon shot in USA (from billions to extinction in 50 years)

1905  *'Theory of Relativity'*, published by Albert Einstein ( $E = mc^2$ )

1905  Anna Pavlova danced *The Dying Swan* at St. Petersburg (Michel Fokine, Saint-Saëns)

1906  Marie Curie professor at University of Paris (Nobel Prize for Radioactivity 1903)

1908  Federal Employers Liability Act introduced corporate liability for negligence in US

1908  'Model T', Henry Ford (first affordable mass-produced motor car)

1909  Cross Channel Flight by Louis Blériot (Calais to Dover in 36½ minutes)

1909  RMS Mauretania took Blue Riband for fastest Atlantic crossing (held for 20 years)

1910  *Journal of Genetics* (William Bateson & R Punnett described animal Inheritance)

1911  Atomic central charge (later termed 'nucleus') proposed by Ernest Rutherford

1911  National Insurance Act (culmination of Liberal reforms incl. pensions, begun 1906)

1911  Roald Amundsen first to reach the South Pole (Scott died 1912 on the way back)

1913  *The Rite of Spring* by Igor Stravinski (Ballet Russe, Paris, Sergei Diaghilev)


CE          The First World War and the Depression  *(CE = years of the current era)*

1914  World Population estimate 1.8 billion (British Empire 450M, *i.e.* one quarter)

1914  First World War triggered by assassination of Archduke Franz Ferdinand of Austria

1915  *The Unconscious*, Sigmund Freud, published (metapsychological paper)

1915  Armenian genocide by Ottoman Turks (1M plus died in forced marches)

1916  Battle of the Somme July–November (1M killed or wounded)

1917  Russian Revolution ('October Revolution', Bolsheviks led by Vladimir Lenin)

1918  First World War ended (Armistice took effect at 11am on 11th November)

1918  Flu Pandemic (H1N1 virus) killed 50M–100M people, ended abruptly Dec 1920

1918  Figidaire company started mass-production of domestic refrigerators

1919  League of Nations set up following Paris Peace Conference & Treaty of Versailles

1919  First Scheduled Passenger Flights (London to Paris)

1922  BBC Radio, first broadcast from London (Station 2LO)

1923  Republic of Turkey established under Mustafa Kemal Atatürk (died 1938)

1923  Population exchange between Greece and Turkey (1.5M Greeks and 0.5M Turks)

1923  Leica Camera, First batch of 31 made by Leitz (invented 1914 by Oscar Barnack)

1925  'Iditarod' 1,085 kms dog sled relay of diptheria antitoxin to Nome in 5½ days

1925  *The Trial*, Franz Kafka, published posthumously (died 1924, aged 40) "Kafkaesque"

1927  *Sein und Zeit* ('Being and Time'), Martin Heidegger, published (died 1976)

1927   World Population reached 2 billion (United Kingdom 45M, London 8M)

1928   Universal Suffrage in United Kingdom (men and women vote on equal terms)

1928   Bactericidal properties of Penicillin noted by Alexander Fleming

1929   Wall Street Stock Market Crash (NYSE Crash, start of ten-year Depression)

1930   Salt March led by Gandhi to challenge British rule (shot by Hindu fanatic 1948)

1930   Purpose-built Odeon Cinema, Welling, Kent (Odeon bought by Rank 1938)

1930   'King Kullen' (Michael J. Cullen) Supermarket opened in Queens, New York City

1931   Electron Microscope demonstrated by Ernst Ruska (later developed by Siemens)

1935   Radio Device for Locating an Aircraft patented by Watson Watt (became RADAR)

1936   BBC Television broadcast from Alexandra Palace (suspended 1939 to 1946)

1936   *Migrant Mother*, Dorothea Lange (photograph for Farm Security Administration)

CE      The Second World War, Before and After   *(CE = years of the current era)*

1937   *Guernica*, painting by Pablo Picasso shown at Paris International Exposition

1938   Kristallnacht, 9th/10th Nov (prelude to the Holocaust by Adolf Hitler's Nazi Party)

1938   Baliem Valley in New Guinea found to hold last untouched stone age society

1939   *Grapes of Wrath*, John Steinbeck (farm tenants in the Great Depression)

1939   Germany invaded Poland and Britain declared war on 3rd September

1940   Battle of Britain June–Sept. (Britain stood alone, Dunkirk, Churchill, RAF Spitfire)

1940   Katyn Massacre (Russians executed 22,000 Polish prisoners, mainly officers)

1941   'Pearl Harbour' (7th December) Japanese air attack on American fleet in Hawaii

1942   Second Battle of El Alamein (23rd Oct to 11th Nov) ended in Allied Victory

1943   German siege of Stalingrad broken by Russians (up to 2M killed or wounded)

1944   Colossus Mk.1 Computer operational, Bletchley Park (Flowers, following Turing)

1944   Gloster Meteor entered service with the RAF (first functional jet engined aircraft)

1944   Normandy Landings, D-Day 6th June (24,000 airborne plus 160,000 amphibious)

1945   Hitler's Suicide and German Surrender ended War in Europe (V-E Day, 8th May)

1945   Auschwitz liberated (5M–6M Jews plus many others killed in the Holocaust)

1945   Atomic Bombs dropped on Hiroshima and Nagasaki (ended Second World War)

1945   United Nations established (headquarters in New York)

1946   Nuremberg Trials, International Military Tribunal (Led to Int. Criminal Court, 1998)

1947   Partition of India and creation of Pakistan (14M displaced, 500,000 died)

1947   AK47 Russian Assault Rifle designed by Mikhail Kalashnikov (75M produced)

1948   National Health Service Introduced (based on Beveridge Report of 1942)

1948   Foundation of the State of Israel (followed Balfour Declaration, 1917)

1948   *Sexual Behaviour in the Human Male*, Alfred Kinsley (*… in the Female*, 1953)

1949   Artificial Insemination of Cattle using frozen semen (pioneered in UK)

1950   Diners Club issued the first Charge Card (American Express Credit Card 1958)

1950   Kidney Transplant in Illinois, USA (the first with living donors were 1954)

1950   Myxomatosis introduced to Australia (UK 1953 via France, 95% Rabbit mortality)

1951   High-density Polyethylene (most common plastic) commercial production

1951   Lyons Corner House LEO, the first office computer (Lyons Electronic Office)

1951   *The Sea Around Us*, Rachel Carson, published (*Silent Spring*, 1962)

1952   De Havilland Comet Jet entered commercial service (London to Johannesburg)

1952   'Great Smog' (air pollution brought London to a standstill, 5th-9th December)

1953   Joseph Stalin died aged 74 (responsible for 20M+ civilian deaths)

1953   Double-Helix DNA proposed by Watson & Crick (earlier X-ray by Rosalind Franklin)

1953   *Philosophical Investigations*, Ludvig Wittgenstein (posthumous, died 1951)

1954   *The Archetypes and the Collective Subconscious*, Carl Jung, written 1934-54

1954   Commercial Transistor (silicon chip semiconductor) produced by Texas Instruments

1954   Brown v Board of Education, US Supreme Court ended school segregation

1954   *Rock around the Clock*, Bill Haley, brought 'Rock and Roll' to popular culture

1956   First Nuclear Power Station came on line (Calder Hall, Cumberland, England)

1957   *Dr. Zhivago*, Boris Pasternak (published by Feltrinelli) Nobel Prize 1958 (died 1960)

1957   Fortran released by IBM, San José (COBOL appeared 1959)

1957   Radio Telescope commissioned at Jodrell Bank observatory (Bernard Lovell)

1958   European Economic Community founded (UK joined 1973, European Union 1993)

1959   M1 Motorway, first section opened from Watford to Rugby (London-Leeds 1977)

1960   "*Wind of Change …*", speech by Harold Macmillan signalled British decolonisation

1960   *To Kill a Mocking Bird*, Harper Lee, published (dealt with racial prejudice)

1960   Lasers demonstrated (Maiman: Pulsed – Ali Javan with others: Continuous)

1960   World Population 3 billion, UK Population 78% Urban, USA 70%, China 16%

1960   Combined Oral Contraceptive Pill ('The Pill') authorised in the USA

1962   Oral Polio Vaccine licensed (dev. Albert Sabin – Injectable by Jonas Salk 1955)

1962   Mao Zedong 'Great Leap Forward' ended (perhaps 30M died in 4 years)

1962   Algeria Independence ended war with France (1M 'Pieds-noirs' emigrated to France)

1962   Royal Ballet *Giselle* danced by Margot Fonteyn with Rudolf Nureyev (asylum 1961)

1962   *One Day in the Life of Ivan Denisovich*, Aleksandr Solzenhitsyn (Nobel Prize 1970)

1963   'Great March on Washington' Martin Luther King (led to Civil Rights Act 1964)

1964   Armed Unmanned Aerial Vehicles (Drones) used in the Vietnam War

1964   Digital Equipment Corporation (DEC) PDP-8 (12-bit Mini Computer)

1965   State Funeral for Winston Churchill (dying days of the British Empire)

1966   Cultural Revolution started in China, Red Guards, continued 10 years (Mao died 1976)

1969   Apollo 11 landed humans on the Moon and returned them to Earth

1969   'Woodstock' Musical Festival (estimated audience 400,000)

1969   TWA Flight 840 with 120 passengers hijacked by Arab Nationalists (Leila Khaled)

1970   Nobel Peace Prize awarded to Norman Borlaug ('Father' of the Green Revolution)

1972   Photograph of girl running naked from US napalm attack (by Huynh Cong Ut, AP)

1973   'Winchester Disk' introduced by IBM (CD-ROM, Sony-Phillips 1985 – DVD 1995)

1973   Magnetic Resonance Image (MRI) first published by Paul Lauterbur (fMRI 1995?)

1974   Rubik's Cube invented by Hungarian sculptor, Professor Ernő Rubik

1975   American defeat and withdrawal from Indochina (end of Vietnam War)

1976   Supersonic airliner Concorde entered Service (LHR to JFK record 2 hrs 53 mins 1996)

1976   *The Selfish Gene*, Richard Dawkins, published (gene-centred view of evolution)

1978   Navstar 1 Global Positioning System (GPS) satellite launched

1978   North Sea Herring fishery collapsed (peak landings were 1.2M tonnes, 1965)

1979   *In Vitro* Fertilisation (IVF) first baby born in Glasgow

1979   Smallpox global eradication certified (World Health Org. programme began 1958)

1980   Ethernet Local Area Network (LAN) protocol, commercial introduction

1980   Dialog database search engine commercially available *via* dial-up modem

1980   Nobel Prize awarded to Frederick Sanger for DNA Genome Sequencing

1981   AIDS first observed in the USA (pandemic peaked in 2005 at 2.2M deaths)

1981   Microsoft MS-DOS became the default operating system for the new IBM PC

1982   IRA Bombed Hyde Park and Regents Park (killed 11 military and 7 horses)

1982   International Whaling Commission adopted a moratorium on commercial whaling

1983   C++ Object Oriented computer language (OOP) released by Bell Laboratories

1983   US Embassy in Bierut destroyed by Islamic Suicide Truck Bomb

1986   Commercial Whaling Moratorium came into force

1986   'Big Bang' Financial deregulation of the City of London

1987   Great Storm along English Channel, then 'Black Monday' Stock Exchange Crash

1988   *The Satanic Verses*, Salman Rushdie (Shia Fatwa ordered Muslims to kill him)

1988   PanAm Flight 103 downed over Lockerby by suitcase bomb (killed 270 people)

| | |
|---|---|
| 1989 | Fall of 'Berlin Wall', 9th November, then Unification of East and West Germany |
| 1990 | Internet Service Providers got commercial access to WWW (TCP/IP standard 1982) |
| 1990 | Hubble Space Telescope launched (smaller precursors by NASA and UK in 1962) |
| 1991 | Digital Cellular Mobile Phone Network (2G, Finland, GSM Standard) |
| 1991 | Collapse of Soviet Union (formal independence of twelve ex Soviet Republics) |
| 1991 | Tomahawk Cruise Missiles used in Gulf War (from surface ships and submarines) |
| 1992 | Grand Banks Cod fishery collapsed (peak landings were 810,000 tonnes, 1968) |
| 1994 | Multi-Racial elections in South Africa ended Apartheid (Nelson Mandela, president) |
| 1994 | Rwandan Tribal Genocide (Hutus killed 800,000 Tutsis and moderates) |
| 1995 | Srebrenica Massacre (8,000 men and boys killed by Serb Army despite UN protection) |
| 1995 | Canon EOS DCS-3 interchangeable lens digital camera (Kodak 1.3 MP Back) |
| 1997 | Northrop Grumman B-2 entered service (Strategic Stealth Bomber) |
| 1997 | Google Search Engine for the World Wide Web developed (Page and Brin) |
| 1998 | Expansion of the Universe found to be accelerating (Reiss *et al*, 'Dark Energy') |

CE        The Start of the Third Millenium …   *(CE = years of the current era)*

| | |
|---|---|
| 2000 | Atmospheric CO2 *ca.* 380 ppm (900 – 1800 CE was constant *ca.* 280 ppm) |
| 2000 | World Population 6 billion, 50% Urban, London 7M, World Prisons 8M |
| 2000 | 'Golden Rice', Genetically Modified for enhanced nutrients (GM Tobacco 1986) |
| 2001 | Creation of Wikipedia (multilingual Web-based encyclopædia) |
| 2001 | September 11 attack on the World Trade Center by Islamic Terrorists (3,000 killed) |

# Epilogue

Now dear reader it is your turn to call the world's events. This chronology ends at the start of the Third Millenium of the current era. The events of this new century are too close to us at the time of writing. It is not yet possible to tell which will resonate in the future.

The past shapes the present and the future. But future events also change our perception of the past. So I hope you will now consider and decide for yourself those 'happenings' in the past that seem most relevant to you and to your world view.

If you end up with your own chronology, however similar or different it may be to this one, then I shall be very pleased.

*MN*
*January 2015*

*Note to Indexes:*

*There are two indexes: This General one for Things, Places and Events, and a second for People. The entries in the indexes give date references. However please note that a date reference is not necessarily the date of an actual event. Date references simply point to the place in the Chronology where an event, place or person is detailed. Date references in the indexes are all in the current era (CE) unless noted otherwise.*

*Note to Indexes:*

*There are two indexes: One for Things, Places and Events, and this second one for People. The entries in the indexes give date references. However please note that a date reference is not necessarily the date of an actual event. Date references simply point to the place in the Chronology where an event, place or person is detailed. Date references in the indexes are all in the current era (CE) unless noted otherwise.*

Notes

# Notes